DWARFS

A Story for Grown-Ups

Once upon a time... OK, we've got that over and done with now. This story, however, is not a fantasy. It is based on many different bits of factual history that, when viewed in isolation, can be discounted, but suggest something far more when brought together. If you want to verify anything, see my three previous books, which cite references wherever possible.

Written by Martin Thomas

&

Illustrated by Carys Pedler

10 9 8 7 6 5 4 3 2 1

Previous Publications by the Author:
The UFO, E.T, Alien Trilogy
January 2026
978-1-83709-292-5

When the Fae Came
December 2025
979-8-25441-733-0

Mythical Monsters, Sunken Continents, and Aliens
March 2026
978-1-83709-417-2

Website Address: MartinThomasAuthor.com

Individual copies of these books can still be acquired from Amazon or Barnes & Noble.

Contents

The Stoneworkers 7

Early Outsider Actions 13

Actions by Dwarfs 17

Dwarfs in Europe 21

The Situation in the Twenty-First Century 27

The Stoneworkers

Back at the end of the Ice Age, the interior of Anatolia, or as we now call it, Turkey, was mainly populated by nomadic hunter-gatherers, and there were some coastal settlements. The first evidence of an inland township dates from 8000 BCE. The dwellings were built of mud bricks with plastered walls and floors.

Göbekli Tepe

This settlement was near to Göbekli Tepe, thought to have been a temple complex, where the bottom layer dates to roughly 9000 BCE.

The township of Çatalhöyük was built in about 7500 BCE, and this was also built of mud bricks, but to a much higher standard. Strangely, all the dwellings were built with very low doors and ceilings, so that a modern-day human would need to crouch the whole time. The entrances to these buildings were from the roof and down ladders.

The townships of Cappadocia date from the eighth and seventh centuries BCE, and differ significantly from earlier settlements in Anatolia in that they were tunneled out of the solid rock. It is estimated that up to 20,000 people could have lived in the dwellings in the underground township of Derinkuyu alone. These were as much as 100 m below ground, but with narrow and low internal corridors and rooms often 1.5 m high or less.

It has been claimed that the 15,000 narrow tunnels were ventilation shafts for the larger tunnels there today. However, these latter shafts were probably created by the much later Christian occupants. It is more likely that the smaller tunnels were the original ones, and that they were cleverly designed to provide both access and ventilation.

A Tunneled Township in Cappadocia

It has also been suggested that the inhabitants must have used child workers for all this quarrying. They would have to be very busy children! It would seem that, a long time ago, there was an influx of skilled stoneworkers who stopped off at Göbekli Tepe and then moved on first to Ḉatalhöyük and then to Cappadocia to set up home. Perhaps some of these stoneworkers also went south, maybe to build the sphinx and to teach the Egyptians how to build pyramids, with the earliest of these dating from 2500 BCE.

Anatolia seems to have been a very popular place to fight over. We've probably all heard of the fall of Troy at the hands of the Greeks in 1190 BCE, but before then there were also empire-builders such as the Kura-Araxes, Hattians, and Hittites.

During all these turbulent times, there was one constant: a mysterious race living in Anatolia called the Dactyls. These had a fearsome reputation as magicians, scientists, healers, and metalworkers. As Arthur C. Clarke said, "Any sufficiently advanced technology is indistinguishable from magic." Put at its simplest, this race of Dactyls was technologically advanced.

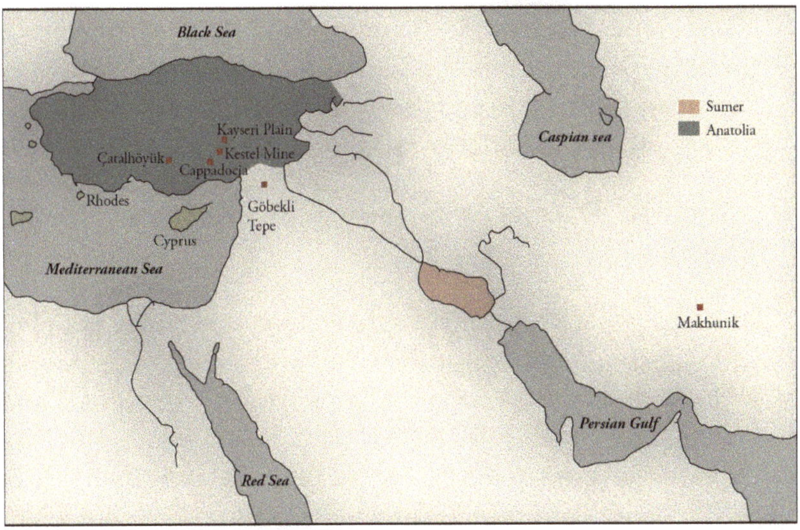

Anatolia and Sumer

These were probably the incomers who had the stoneworking skills to create Göbekli Tepe structures and teach the Egyptians how to build pyramids. It is interesting that similar capabilities were being developed in Mesoamerica and Southeast Asia in the same period. This does not mean that the Dactyls themselves were present in those places too, however.

Dactyls were reported as resident in Crete and other Greek islands for so long that they have been absorbed into Greek mythology. They apparently predated the birth of Zeus, which has been estimated to have been about 4000 BCE. They were described as mountain dwellers, and it is claimed they were the first to discover how to smelt iron ore. By about 400 BCE, Greek writers considered them to have been real people.

We cannot identify these Dactyls until we look at Greek villages where Dactyls are reputed to have lived:

> **Phrygian Dactyls** near Mount Ida in Crete.
>
> **Kabeiroi Dactyls** on Lemnos and Samothrace, and in Thebes. Pliny the Elder, in his book *Natural History*, referred to the little people of Thebes, who were called the Catizi Dwarfs.
>
> **Rhodian Dactyls**, also known as Telchines, on Rhodes.

The occupants of these villages were called "little people," with female Dactyls being called Hekaterides. These people were famed metalsmiths, stoneworkers, healers, and magicians. THEY WERE DWARFS.

These Dwarfs were described as typically less than 1.5 m tall, with generally human proportions, although with slightly larger heads and very broad shoulders. The men had bushy beards reaching halfway down their chests.

It is possible that these Dwarfs were assisted by other outsiders who looked the same as humans but may not have been genetically the same. We'll come back to them later on.

Thus, we have a group of outsiders living in isolation, working for the good of the humans around them. Does that sound familiar? Not for nothing have people claimed that Anatolia was the location of Atlantis, despite the fact that it was not outside the Pillars of Hercules (probably the Straits of Gibraltar) and it wasn't surrounded with canals or on an island.

Early Outsider Actions

It all started when humans progressed from troops of nomadic hunter-gatherers to become farmers. How did this happen? The simple answer is that we were taught. Originally there were three areas where these teachers were based—the Caribbean, the now-flooded

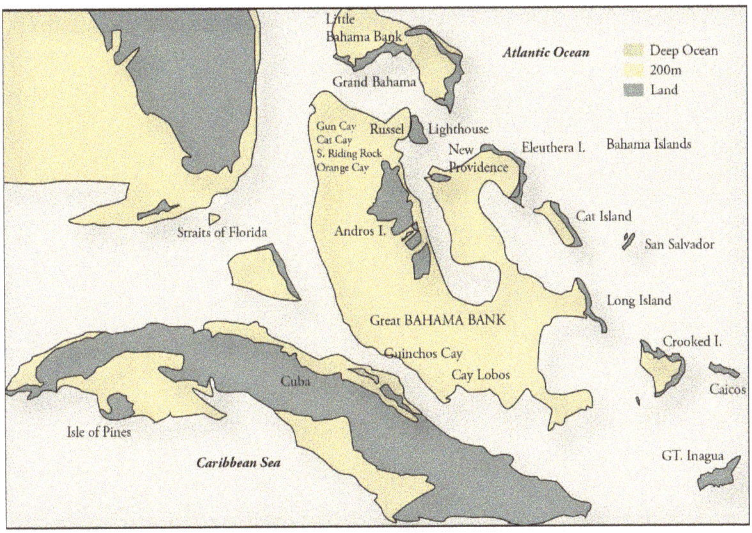

Cuba and the Bahama Banks

area of Sundaland (Mu) in Southeast Asia, and the Middle Eastern areas of Mesopotamia and Turkey. Sadly one of these sites, the one in the Caribbean, was destroyed about 10,000 years ago by the impact of a comet in the nearby ocean, but the others thrived, despite setbacks, leading to civilizations in China and Europe.

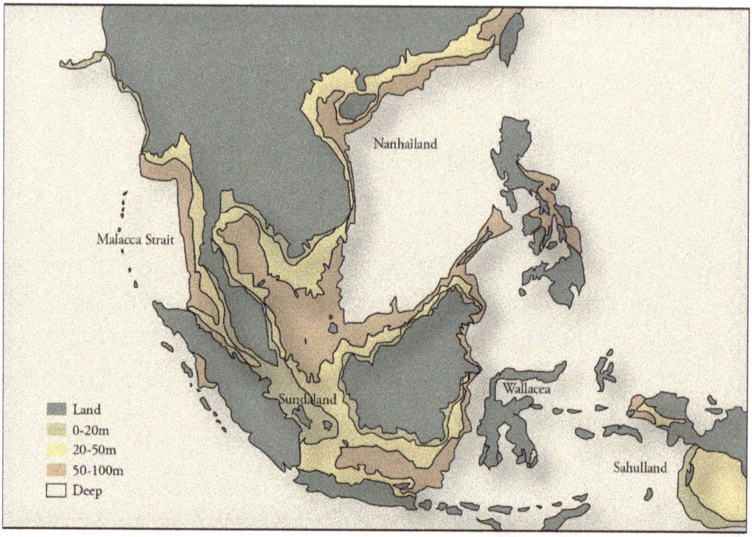

Sundaland (Mu) – Southeast Asia

It was the Caribbean base which gave rise to the legend of Atlantis. Its location was surrounded by the sea, and its destruction did occur in a cataclysm, unlike other proposed sites which were gently swallowed up as the sea level rose at about 5 cm per year. The teachers' base there was probably on Cuba, while the surrounding lands of Mesoamerica and the now-submerged Bahamas Banks were occupied by their students.

This begs the obvious question: who were the teachers? The answer is equally obvious. It had to be another species, more developed than our own, either already underground or underwater here on Earth, or from somewhere else either in this universe or even another hypothetical parallel universe.

Applying Occam's razor, I prefer the simplest solution. With millions of planets around just our local galaxy, many thousands will have their own civilizations, and hundreds of these will be more advanced than ours. Some could have discovered how to navigate the vast distances of space, and these probably would have found us millennia ago. A few must have decided to raise human moral and technological standards—the popular term in science fiction is "Uplift."

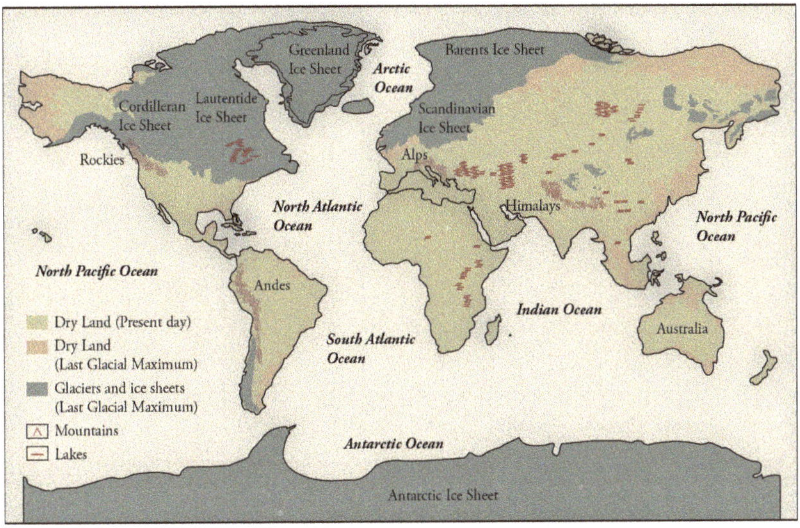

Ice Sheets at the Last Ice Age Maximum

So when did these outsiders arrive? As best as we can determine, they came here some time before the end of the last Ice Age, which was over 10,000 years ago. This Ice Age did not envelop the entire Earth.

There were huge ice sheets across Northern Europe, America, and Asia, and temperatures worldwide were lower, but the equatorial regions were still habitable.

These areas included Sumer in Mesopotamia and probably the Black Sea basin, where the water level was much lower in those days. In these areas, a group of humans were taught agriculture, mathematics, writing, schooling, and the calendar by a species described in clay tablets found by archaeologists in Sumer as over 2.5 m tall.

Agriculture spread out around the world, giving humans more time for food production and, sadly, for fighting. We were then taught more in Anatolia when Dwarfs took over, showing us how to further manipulate our environment.

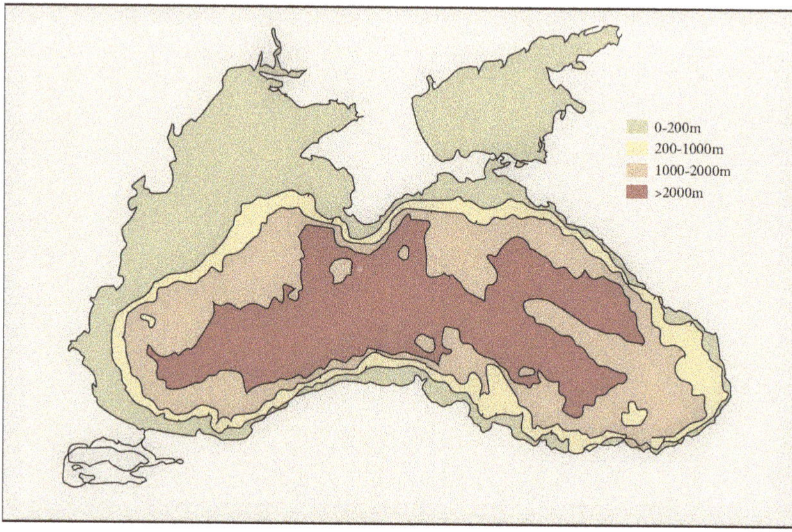

The Black Sea

The Black Sea basin was about 120 m below current sea levels back in 10000 BCE, and it was probably the inhabitants of this area who were the subjects of the Dwarfs' efforts. They were early recipients of the knowledge given to us in Sumer, and their lands were very fertile, so they had a thriving agricultural base.

Actions by Dwarfs

So what did the Dwarfs do to help humans? This changed over time. During the late Stone Age, in their first diaspora, some of the Dwarfs spread across Europe, advising, helping, and perhaps even building the stone ritual structures including La Torre-La Janera in Huelva in Spain (5500 BCE), Stonehenge in Britain (3000 to 2500 BCE), and the Knap of Howar on Papa Westray in Orkney Scotland (3700 BCE).

A Neolithic Tomb

A lot has been written about the fact that the stone used was often granite, which contains piezoelectric crystals of quartz, but we do not have the evidence to suggest any ulterior motive other than it being a very hard rock.

There is plenty of archaeological evidence that the society living in the Black Sea basin was an early recipient of the Dwarfs' stoneworking education, with signs of stone ports, public buildings, and wooden structures beneath the waters. They were apparently capable coppersmiths.

During the Copper Age, the Dwarfs taught us mining techniques to help us obtain copper, as well as the skills for working with it. There were good supplies in 7500 BCE in Anatolia and also nearby in Cyprus.

The Sea Breaks into the Black Sea

However, in about 6000 BCE, the waters of the Mediterranean Sea broke through into the Black Sea, raising its water level by about 120 m. This, of course, destroyed the budding civilization as the inhabitants were forced to scatter to all points of the compass. Those heading south ended up in Anatolia, where they moved into the township of Çatalhöyük, which had been abandoned by the Dwarfs when they moved into Cappadocia.

They were probably joined by other humans moving into the Anatolian interior, building up the population, and applying their knowledge of agriculture to new areas.

Things changed drastically, however, when the Dwarfs eventually felt it was time to pass on their knowledge of how to use the tin they had found in Kestel Mine, north of Cappadocia. Further north from there, on the Plain of Kayseri, the Dwarfs had their workshops, and it was here in about 3000 BCE that they first showed us humans how to make bronze by adding the tin to copper, and changed the world for ever. The Bronze Age had begun.

When the supplies of tin in Anatolia started to run low, this started the second diaspora of Dwarfs across Europe, teaching humans how to mine copper and tin and how to make bronze. Bronze Age civilizations, such as the Minoans on Crete, rose and fell.

Bronze Monastery Bell

Last, in Anatolia at least, in about 1300 BCE, came the making of iron by smelting iron ore.

This led to the third and final diaspora, as the Dwarfs and all their companions, having taught the human inhabitants of Anatolia how to make iron, abandoned their dwellings and moved out, west to Spain and Portugal, north to Norway and Sweden, and east to India.

On the border between Iran and Afghanistan is another township, Makhunik, where, like Ḉatalhöyük, the houses have very low doorways and ceilings. The significant difference is that within the last hundred years, there was a race living there who were less than 1.5 m tall. We don't know whether this township was built by the stoneworkers on their way to Göbekli Tepe before 9000 BCE or on their way toward India after 1300 BCE. Claims of its age vary between 1,500 years and 5,000 years, so the latter date would seem the more likely.

In 280 BCE, the inhabitants of Rhodes erected a statue that was considered one of the seven ancient wonders of the world—the Colossus of Rhodes. This was 32 m high and stood in their harbor, although not astride the entrance. It was created by a Rhodian sculptor called Chares of Lindos, and was made of bronze reinforced with iron rods and stone. Chares was not a Dwarf, but he had inherited the legacy of metalworking from the Rhodian Dactyls of the past.

Dwarfs in Europe

In Germany and Austria there are tales of Dwarfs with very musical voices living in the Black Forest. It was claimed that their magical powers came from their red hats or helmets, and they became powerless if they lost them. Perhaps they had their technical marvels built into these hats. They were described as having very pale skin and small hands and feet.

At about this time, it was recognized that there was more than one type of Dwarf. Perhaps different groups had begun to break away from the norm, or perhaps they had been a diverse band when they first arrived. Variations were:

A Gnome

A pale Dwarf, 1.5 m tall, constituting the majority

A hairy Dwarf, 1.3 m tall, living in mountains above the snowline

A brown Dwarf, 0.6 m tall, called a Duegar, that seemed to like causing trouble in houses

A tiny Dwarf, 0.3 m tall, better known as a Gnome. These were generally friendly, and may have been the source of the term Knockers or Tommy-Knockers, because they liked to live in mines and would warn of danger by knocking.

Perhaps these last two are the Dwarf equivalents of the rebellious teenager and the precocious brat. I suppose the hairy Dwarf could have been a misidentified juvenile Yeti.

One writer claims there are also black, red, and green Dwarfs, but until I get independent corroboration I'm going to file these together with the paisley one and the blue one with orange spots.

In the Middle Ages, the majority of humans believed that Dwarfs existed, so there are many tales of Dwarfs. We will never know, though, whether individual reports were of humans with dwarfism, or of our technological Dwarfs. It has even been suggested that, in the distant past, humans and Dwarfs interbred to some extent, and that modern human dwarfism may be a consequence of this, occurring randomly many generations later.

There are tales of the persecution of "little people" all over the world, inspired by the Church, which considered them demons. They were often enslaved or hunted down and killed. It seems unlikely that these "little people" could have been Dwarfs, because they were quite capable of disappearing or defending themselves.

Once the inhabitants of iron-rich areas had learned how to smelt iron, there was little more that the Dwarfs could do other than to gently help in the Earth's technical development, small steps at a time.

They encouraged the development of canals from early examples in China in the tenth century until Leonardo da Vinci designed the first double miter pound lock still used today. It has often been suggested that Leonardo's futuristic technical designs were so wide-ranging and numerous that he must have been influenced by an outside source—Dwarfs maybe?

The Age of Canals

This led to the first Industrial Revolution in Britain, which began in the 1750s CE when canals were built to move the coal and iron, and when cotton and wool weaving machines were invented.

The wrought-iron bridge over the River Severn was a masterpiece of the period.

The Dwarfs surely must have helped in the second British Industrial Revolution, which began in the early 1800s with the start of the building of railways, culminating in what was known as "Railway Mania" in about 1850.

Ironbridge

Among the famous engineers of this period—Thomas Telford, Abraham Darby, Robert Stephenson, and James Watt—Isambard Kingdom Brunel stood out, even though he was only 1.5 m tall. Who is to say whether he was a human/Dwarf hybrid throwback?

Isambard Kingdom Brunel

The Age of Steam

We do know that Dwarfs moved around the world to the United States of America and the USSR. We don't know where they were based in the USSR, but in the USA they set up a base near Pittsburgh, probably under a lake there. There were a number of Dwarf sightings in that area in the first half of the twentieth century.

An Underwater Base

There was one more technical advance to be made: the Bessemer Process for making steel. In 1856, in Sheffield, England, Sir Henry Bessemer said it came to him in a dream. Unless the Dwarfs can also project their thoughts as one of their high technology capabilities, there may have been another non-human entity involved here.

There are also a couple of reports of Dwarfs in Venezuela getting involved in fights with locals. Apparently they are incredibly strong, with skin as tough as old leather.

The Situation in the Twenty-First Century

In the late nineteenth and early twentieth centuries, the Dwarfs started looking to their future in the belief that any further education of humans would require skills other than theirs.

It is known that in 1941 three Dwarfs approached the King of Sweden, apparently bypassing all his security. Their conversation is not reported, but it was almost certainly a request for permission to build a colony underground in northern Sweden. There is evidence to suggest this was granted and that the colony was constructed.

An Unidentified Submerged Object

It is not known whether the Italian government was approached in the same way, but back in the early twentieth century, vibrations could be detected below the city of Genoa, with increased sightings of Dwarfs. It is thought that their colony there has an underwater entrance.

It has been suggested that the Dwarfs' human-looking helpers have also set up home. Those from Anatolia and perhaps Sundaland in Southeast Asia, as well as any survivors from the Caribbean, have had many thousands of years to fade into the background. It is claimed that they are the modern-day Basques and, as proof, the proponents of this idea claim that the Basques are very similar to the Mayans, who could once have been helping the Atlantean teachers.

Basque

Mayan

God-King Pakal

Apparently, Basques and Mayans have similar aquiline profiles, much like the profile of the ancient Maya God-King Pakal. They also use similar agricultural methods and play a similar ball game, known as Pelote in Europe. The Basque language is like a remote Siberian language (the descendants of the helpers from Sundaland?) and the language of the Indio tribe of the Otomi. This tribe actually predates the Aztec tribe, and has remained in total isolation for thousands of years.

Basques possess the world's highest frequency of Rhesus (Rh) negative blood and a high prevalence of type O blood. Their unique genetic profile, likely stemming from long-term isolation, includes roughly 25–35% Rh-negative frequency and a very low prevalence of the B gene. The Otomi, on the other hand, while again having a very high prevalence of type O blood, are almost entirely Rhesus positive.

It is difficult to explain this major discrepancy unless they were different races or species. However, they both built pyramids, and some Mayans claim that they are directly descended from the sky-people.

Progressively the Dwarfs have withdrawn from around the world and taken up residence in their two known colonies in Sweden and Italy although, of course, there may be more. The only place now where Dwarfs can still be seen is in South America, where there are still many mines awaiting development and the local inhabitants still need help avoiding the big mining corporations sweeping them aside.

This does not mean that Dwarfs have withdrawn entirely from contact with humans. Their two known colonies are close to the major steel-producing areas of Sweden and Italy, and there is perhaps another near Pittsburgh. In addition, they do seem to be performing underground operations in other parts of the world, although we don't know on whose behalf they are working.

Hollywood Dwarf

Historic Dwarf

Our current image of Dwarfs is mainly dictated by cinema and the games industry. Together they have given us the idea that Dwarfs are solely warriors, whereas they are really so much more. Dwarfs have played a major role in the development of humankind; they certainly deserve our gratitude, and they have earned the right of residence on Earth. It would be interesting to know who has now taken over their role as our hidden teachers.

The End

OTHER BOOKS BY MARTIN THOMAS

UFO - FRIEND OR FOE?

ABOUT THIS BOOK

Millions of people believe that there are mysterious objects flying in the sky. They were originally described as Unidentified Flying Objects—UFOs—but have more recently been termed UAPs—Unidentified Anomalous Phenomena. There are many more sightings of UAPs over land than have been associated with maritime UFOs, which have been termed Unidentified Submersible Objects (USOs). The abbreviation UAP is now used to cover both UFOs and USOs.

UAPs seem to fall mainly into two distinct categories. They may be objects which are apparently metallic in construction, of many shapes and varying in size from larger than a car to absolutely massive. Alternatively they can be balls, either of some form of plasma, or metallic, giving off or reflecting light, and varying from the size of a tennis ball upward.

This book looks at UFO sightings worldwide and at the various options for their origin. It looks at the possibility of alien bases, at their frequency and distribution, and their likely purpose. (90 pages)

WE OWN 29%, ET HAS THE REST

ABOUT THIS BOOK

About 30% of Earth is land, and we know very little about what is under the sea. Even our navies can only get a glimpse, and they aren't telling us everything.

The galaxy is a cruel place, with some species looking on others as sources for slaves and even as food. For some, survival is a challenge to be addressed. The solution adopted by many is to send out small self-sustaining colonies that can hide away. If something happens to their parent planet, at least the species has a chance of survival.

A number of species appear to have chosen Earth as one of their secret colony planets, and have moved in. The trouble is that they didn't tell us or even ask us. Our governments probably know what has gone on, but they don't intend to tell us, so we will have to find out for ourselves. This book makes a start at working out who is where, and who is doing what. (124 pages)

EARTH'S ALIEN SYLLABUS

This is the third part of a trilogy. Having established that aliens exist, and that they are colonizing this planet, I now look at how some of them are helping us advance toward the time when we might be considered acceptable compatriots for other spacefarers.

First they adjusted our DNA to increase our intelligence, and since then they have progressively taught us the technology we will require. They started with agriculture, mathematics, astronomy, and schooling and then taught us how to build monumental structures. Then they moved on to mining, metallurgy, and metalworking, taking us from the Stone Age through the Bronze Age and into the Iron Age. Thereafter they had to be more subtle, using telepathy to promote chemistry, electrical power generation and transmission, and radio telegraphy.

Inevitably, they came up against our baser nature, having to contend with the greed of the rich when they tried to stifle developments that would detract from their power or wealth, and to restrict the knowledge to a privileged few.

Our mentors seem to have bypassed this by encouraging nations to retro-engineer their "crashed" atmospheric craft, making the technology available worldwide again. We wait to see what their next move will be. (74 pages)

THE UFO, ET, ALIEN TRILOGY

This book contains the previously published volumes:

- UFO – Friend or Foe
- We Have 29% – ET Has the Rest
- Earth's Alien Syllabus

In these, I survey the records of UFO and alien sightings, first to try to determine whether UFOs really exist. To simplify my task, I excluded all sightings where there was only one witness, and where the witness could easily have mistakenly reported a sighting of the planet Venus.

I was then able to convince myself easily that not only were UFOs and aliens real, but that they had multiple colonies hidden on Earth either underground or under the seas. By examining the reports, particularly of coastal sightings, I was able to identify nine permanent colonies, although there must be many more.

I was also able to show how a group of aliens were developing humankind progressively, first enhancing our learning potential and then teaching us various aspects of technology, so that we will one day be able to join their community. This they are achieving despite our best efforts. (314 pages)

WHEN THE FAE CAME

ABOUT THIS BOOK

Once I had convinced myself that UAPs and aliens were real, I started looking at where they lived on Earth and why they were here. It became clear that there were a number of alien colonies concealed here underground or underwater. What surprised me was that UAP researchers were interested in large alien creatures, but seemed to assume that smaller ones were really magical beings, and should be avoided.

My early research showed that these little people, or some of them at least, were also aliens living in colonies in Mexico, Malaysia, Canada, and Northern Italy. There was no difference between them and taller species of aliens. They possessed the same technology that could make them invisible, fly, and send and receive thoughts telepathically. I looked at the types of aliens sighted in the British Isles—Fairies, Dwarfs, Elves, Pixies, and Goblins, and their equivalents around the world. Elves are just about everywhere, and Dwarfs appear to have been among us for over 5,000 years, teaching us.

I then looked at worldwide sightings of aliens other than these, and managed to identify at least twenty other races of little people, based on only a small sample of the total number of sighting records available. I'm sure I missed some, or possibly a lot. (105 pages)

MYTHICAL MONSTERS, SUNKEN CONTINENTS AND ALIENS

ABOUT THIS BOOK

This book looks at some of the more peripheral aspects of unusual phenomena which have been perhaps witnessed or remembered in folklore, but never proven.

For a long time there have been stories of reptile people inhabiting caverns beneath Earth's surface. It is shown here that these stories are credible, and the locations of some of their hideaways have been identified.

It feels as though there have always been stories of ghosts and poltergeists, but it has proved possible to show that these are all a consequence of one of two factors, the "Stone Tape Theory" or the mischievous actions of some of our long-time resident aliens—the Fae.

The existence of sea monsters has always been denied, but one after another they have been found. First came the giant squid, then the colossal squid, and then, most excitingly, the existence of Nessie has been proven, not as a lonely inhabitant of Loch Ness, but as an ocean-wide phenomenon.

It has been shown that of the three claimed lost continents, at least Mu can be identified and is now known as Sundaland.

Atlantis is likely to have existed in the distant past, but there are too many possible locations to pick from. Lemuria is very unlikely to have existed as a real place in the Indian Ocean.

However, in practice, Atlantis is an idea rather than a reality. It is probably one of a series of bases used by aliens while they attempt to raise the morality and technical ability of humans. By hiding them in places which would be flooded at the end of the Ice Age, they made certain that humanity would not discover their ruins for some time. (125 pages)

www.ingramcontent.com/pod-product-compliance
Lightning Source LLC
LaVergne TN
LVHW051051210926
848486LV00002B/67
* 9 7 8 1 8 3 7 0 9 4 8 6 8 *